My Mom the Mayor

Printed in Mexico

ISBN-13: 978-0-15-351868-3
ISBN-10: 0-15-351868-5

1 2 3 4 5 6 7 8 9 10   050   11 10 09 08 07 06

Harcourt
SCHOOL PUBLISHERS

Visit *The Learning Site!* www.harcourtschool.com

Morning Meeting

My mom is our mayor.
Today I will visit her at work.

2

Events in Town

Next, we go to a new park.
Mom cuts a ribbon. I can play.

4

Then we help with a race. The flag is a symbol of our town.

5

Ending the Day

At night Mom gives a speech. She talks about jobs. People listen to her. Then they clap.

6

We had a busy day.
Now it is time to go home.

Think and Respond

1. What is Mom's job?

2. Why does Mom cut the ribbon?

3. Why do people clap after Mom's speech?

Activity

Draw a picture of a leader in your community.